Grains

HEALTHY ME

Published by Smart Apple Media
1980 Lookout Drive, North Mankato, Minnesota 56003

PHOTOGRAPHS BY Richard Cummins, The Image Finders (Mark E. Gibson), Bonnie Sue Rauch,
Tom Stack & Associates (Inga Spence), Unicorn Stock Photography (Doug Adams, Nancy P. Alexander,
Eric R. Berndt, Martin R. Jones, Arni Katz, Patti McConville, Bill McMackins, Marshall R. Prescott, Larry Stanley)
DESIGN BY Evansday Design

Library of Congress Cataloging-in-Publication Data
Kalz, Jill.
Grains / by Jill Kalz.
p. cm. — (Healthy me)
Summary: Describes various grain products and their role in human nutrition.
Includes instructions for making crunchy cold bananas.
ISBN 1-58340-301-9
1. Cereals as food—Juvenile literature. 2. Cookery (Cereals)—Juvenile literature.
3. Grain—Juvenile literature. [1. Grain. 2. Nutrition.] I. Title.

TX393.K35 2003
641.3'31—dc21   2002030903

First Edition

9 8 7 6 5 4 3 2 1

Grains

Fields of Food

Grains are special kinds of grasses. Farmers grow grains to feed farm animals and people. There are eight kinds of grains. Farm animals feed on five of them: corn, oats, barley, millet, and sorghum. People eat mostly the other three: wheat, rice, and rye.

Farmers **sow** grains in dry fields, except rice. Rice is planted in **shallow** water. All young grains look like green grass. Then they grow taller and make seeds. When grains are fully grown, they turn yellow.

Grains that are green are still growing. ⌃

5

Oats are grown mostly for animals to eat.

Most farmers **harvest** grains with big machines. The machines cut and dry the grains. Then they shake the grains so the seeds fall off. The seeds are the part you eat.

More than half of all the corn in the world
comes from the United States.

⌄

Machines help farmers harvest grains.

Made with Grains

Wheat and rice are the most common grains. Wheat seeds are crushed into powder. The powder is called flour. Flour is used to make breads and **pasta**. Cakes are also made with flour. So is pizza crust.

Almost all of the rice in the world comes from Asia. Rice is not made into flour. Rice seeds are eaten whole. Millions of people eat rice every day.

People in Asia grow the most rice.

9

Wheat seeds are crushed to make flour.

Many different foods are made with grains. Wheat, rice, corn, and oats are used to make cereal. Some people bake rye bread. Others make sorghum pancakes.

Corn seeds grow in bunches called cobs.
Cobs are covered with leaves called husks.

Harvested grains are kept inside tall
buildings to keep mice and bugs out.

The Good Stuff

Grains give you energy. They are your body's fuel, like gas is a car's fuel. Without gas, a car does not run. Without grains, your body is too tired to move.

Grains have a lot of fiber, **vitamins**, and protein.

Fiber moves food through your body. The vitamins

in grains keep your brain and heart healthy. 13

Protein builds strong muscles.

Sorghum seeds have a lot of fiber. ︿

‹ Grains help your body grow and feel good.

Not all foods made with grains are good for you.

Some foods have sugar added. Most cookies have

sugar. Too much sugar can be bad for you. Eating

brown bread, pasta, and rice is best.

A tortilla is a flat, Mexican bread made from flour, water, and salt.

⌄

Most cookies have very few vitamins.

Eating Right

All foods belong to one of five food groups. Grains belong to the grains group. Foods made from milk belong to the dairy group. There are also groups for fruits, vegetables, and meats.

Your body needs a lot of grains. Doctors say you should eat 6 to 11 helpings of grain each day. A helping may be one slice of bread. A small plate of spaghetti. Or a bowl of cereal.

Spaghetti gives your body lots of energy. ⌃

< Your body needs all kinds of foods.

It is important to eat foods from all of the food

groups. Each group has things your body needs.

Grains are the fuel you need to run, jump, and

think fast!

Dr. John Kellogg made the first cold cereal in 1894. He made wheat flakes.

Most people do not eat barley.
But some grown-ups drink it!
Beer is made from barley.

Crunchy Cold Bananas

Oat cereal makes these bananas crunchy. Oats belong to the grains group.

WHAT YOU NEED

A plate covered with waxed paper
Three bananas
Six popsicle sticks
Two six-ounce (180 ml) containers of yogurt, any flavor
Two cups (550 ml) toasted oat cereal, crushed

WHAT YOU DO

1. Peel the bananas. Break each banana in half.
2. Push one popsicle stick into each banana half.
3. Dip each banana into the yogurt. Then roll it in the cereal and put it on the plate.
4. Put the plate in the freezer.
5. Chill the bananas for an hour. Then eat them!

WORDS TO KNOW

22

harvest to pick or gather food

pasta flour paste that is shaped and dried; macaroni and spaghetti noodles are kinds of pasta

shallow not deep; the ocean is deep, but a puddle is shallow

sow to plant seeds for growing

vitamins things in food that keep your body healthy and growing

Read More

Aliki. *Corn Is Maize: The Gift of the Indians*. New York: HarperCollins Children's Books, 1996.

Egan, Robert. *From Wheat to Pasta*. New York: Scholastic Library Publishing, 1997.

Morris, Ann. *Bread, Bread, Bread*. New York: William Morrow & Co., 1993.

Explore the Web

CHEERIOS.COM

http://www.cheerios.com

KELLOGG'S CEREALS

http://www.kelloggs.com

RICE ROMP: U.S. RICE PRODUCERS

http://www.riceromp.com

Pasta noodles are made from grains.

INDEX

bread 8, 10, 14, 15

cereal 10, 19, 20

corn 4, 7, 11

energy 12

fiber 13

food groups 16, 18

protein 13

rice 4, 5, 8, 9, 14

vitamins 13

wheat 4, 8